Table of Contents

How To Determine Stuck Depth

Definition of Stuck Pipe

Stuck pipe is a situation when the drill string cannot be moved from the well. The pipe may be partially moved and you may be able to circulate and rotate the pipe.

Personnel on the rig must be able to identify the cause of stuck pipe in order to figure out the right way to free the pipe.

Stuck Pipe Categories

1. **There are 3 categories of stuck pipes as follows:**

   Pack off and bridging: Pack off and bridging are occurred when there is something in the wellbore as formation cutting, junk, etc accumulating around drilling string/BHA and that stuff blocks the annulus between drill string and the wellbore. You should remember that either big or small debris can stick the pipe.

   According to statistics around the world, pack off and bridging is the most frequent cause of stuck pipe situation in the world. It normally occurs when the mud pumps are off for an extended period of time such as when pulling out of hole. It is quite a tough job to free the pipe in case of packoff or bridging and the chance of success is lower than differential or wellbore geometry sticking mechanism.

2. **Differential sticking:** Differential sticking happens when drill string is pushed against permeable formations by differential pressure between hydrostatic and formation pressure. The frictional force between drillstring and formation is so high that you will not be able to move the pipe. The differential sticking tends to easily happen when drilling through depleted reservoir is conducted. Moreover, this stuck mechanism almost always happens when the drill string has been

stopped moving for a long time.

3. **Wellbore geometry:** Wellbore geometry stuck pipe mechanism occurs when the shape of the well and the bottom hole assembly (BHA) don't match each other. Therefore, the drill string is not able to pass through that section.

Stuck Pipe Caused by Pack off and Bridging

There are seven cases in this category.

1. Cutting settling in vertical and deviated wells

2. Shale instability

3. Unconsolidated formations

4. Fractured formations

5. Cement block

6. Soft cement

7. Junk

Cutting Settling in a Vertical or Near Vertical Wellbore Causes Stuck Pipe

How does it happen?

Typically, the well classified as a vertical or near vertical well has inclination less than 35 degree. Cuttings in the wellbore are not removed from the annulus enough because there is not enough cutting velocity in and/or mud properties in the wellbore is bad. When pumps are off, cuttings fall down due to gravitational force and pack and annulus. Finally, it results in stuck pipe.

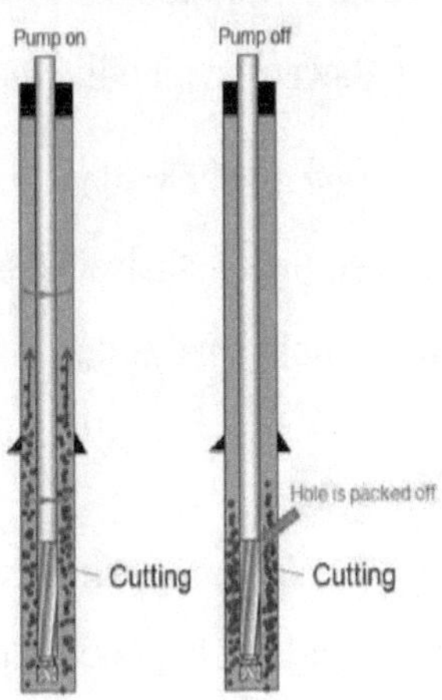

Note: In order to clean annulus effectively, the annular velocity must be more than cutting slip velocity in dynamic condition. Moreover, mud properties must be able to carry cutting when pumps on and suspend cutting when pumps off.

Warning signs of cutting setting in vertical wells

- There are increase in torque /drag and pump pressure.
- You may see over pull when picking up and pump pressure required to break circulating is higher without any parameters changes.

Indications when you are stuck due to cutting bed in vertical wells

- When this stuck pipe caused by cutting settling is happened, circulation is restricted and sometimes impossible. It most likely happens when pump off (making connection) or tripping in/out of hole.

What should you do for this situation?

1. Attempt to circulate with low pressure (300-400 psi). Do not use high pump pressure because the annulus will be packed harder and you will not be able to free the pipe anymore.

2. Apply maximum allowable torque and jar down with maximum trip load. Do not try to jar up because you will create worse situate.

3. Attempt until the pipe is free, then circulate and work pipe until the wellbore is clean. Check cutting at shale shakers, torque/drag and pump pressure in order to ensure hole condition.

Preventive actions

1. Ensure that annular velocity is more than cutting slip velocity.
2. Ensure that mud properties are in good shape.

3. Consider pump hi-vis pill. You may try weighted or unweighted and see which one gives you the best cutting removal capability.

4. If you pump sweep, ensure that sweep must be return to surface before making any connection. For a good drilling practice, you should not have more than one pill in the wellbore.

5. Circulate hole clean prior to tripping out of hole. Ensure that you have good reciprocation while circulating.

6. Circulate 5-10 minutes before making another connection in order to clear cutting around BHA.

7. Record drilling parameters and observe trend changes frequently.

8. Maximize ROP and hole cleaning.

Cutting Settling in deviated wells Causes Stuck Pipe

Typically a well which has inclination more than 35 degree is classified as a deviated well.

How does it happen?

For the deviated well, cuttings tend to set at the low side of the wellbore and form a cutting bed.

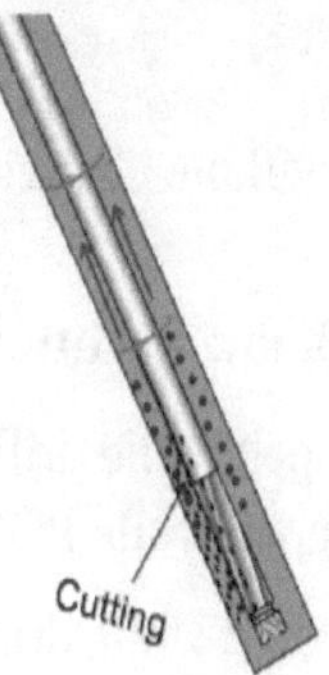

When there is a lot of cutting bed, it will slide down and pack the string. Moreover, while pulling out of hole, BHA will move some cutting bed and finally pack BHA and drill string.

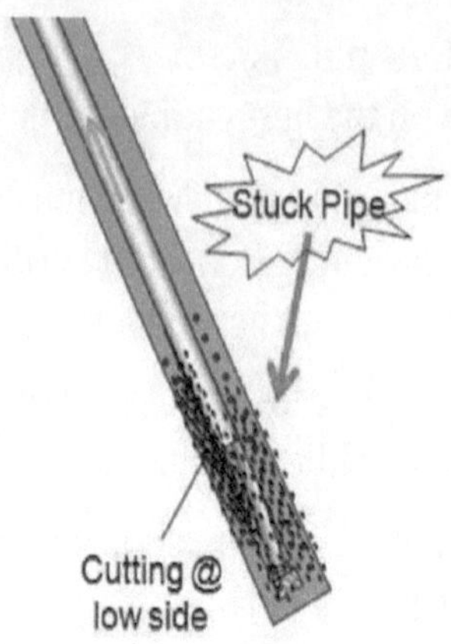

Warning signs of cutting setting in deviated wells

- Drilling with high angle well (more than 35 degree).

- While drilling with a mud motor, cutting cannot be effectively removed due to no pipe rotation.

- Increase in torque and drag (you must have a trend to see if torque/drag is abnormal)

- Increase in pump pressure without changing any mud properties.

Indications when you are stuck due to cutting bed in deviated wells

· The stuck pipe can happen while drilling and tripping out of hole. Most of the time, it will happen while POOH.

· Increase in torque and drag while drilling.

· Increase in drag while tripping out.

· Circulation pressure is higher than normal. Sometimes, it is impossible to circulate.

What should you do for this situation?

1. Attempt to circulate with low pressure (300-400 psi). Do not use high pump pressure because the annulus will be packed harder and you will

not be able to free the pipe anymore.

2. Apply maximum allowable torque and jar down with maximum trip load. Do not try to jar up because you will create worse situate.

3. Be patient, and attempt until the pipe is free, then circulate and work pipe until the wellbore is clean. Do not continue operation until the hole is properly clean. Check cutting at shale shakers, torque/drag and pump pressure in order to ensure hole condition.

Preventive actions:

1. Ensure that annular velocity is more than cutting slip velocity.

2. Ensure that mud properties are in good shape.

3. Consider pump hi-vis pill. You may try weighted or unweighted and see which one gives you the best cutting removal capability.

4. If you pump sweep, ensure that sweep must be return to surface before making any connection. For a good drilling practice, you should not have more than one pill in the wellbore.

5. Circulate hole clean prior to tripping out of hole. Ensure that you have good reciprocation while circulating.

6. Circulate 5-10 minutes before making another connection in order to clear cutting around BHA.

7. Record drilling parameters and observe trend changes frequently.

8. Maximize ROP and hole cleaning.

Shale Instability Causes Stuck Pipe

Shale instability happens when shale formation becomes unstable and finally formations break apart and fall into an annulus.

How it happen?

Water in the mud absorbed by shale formations causes swelling effect on formations. When there is a lot of water, shale will not be able to hold their particles together and finally falls apart into the well. Finally shale particles will jam a drill string.

The shale instability is a chemical reaction which is time dependent. It means that you may not see it on day one, you may see it after you have been drilling for days.

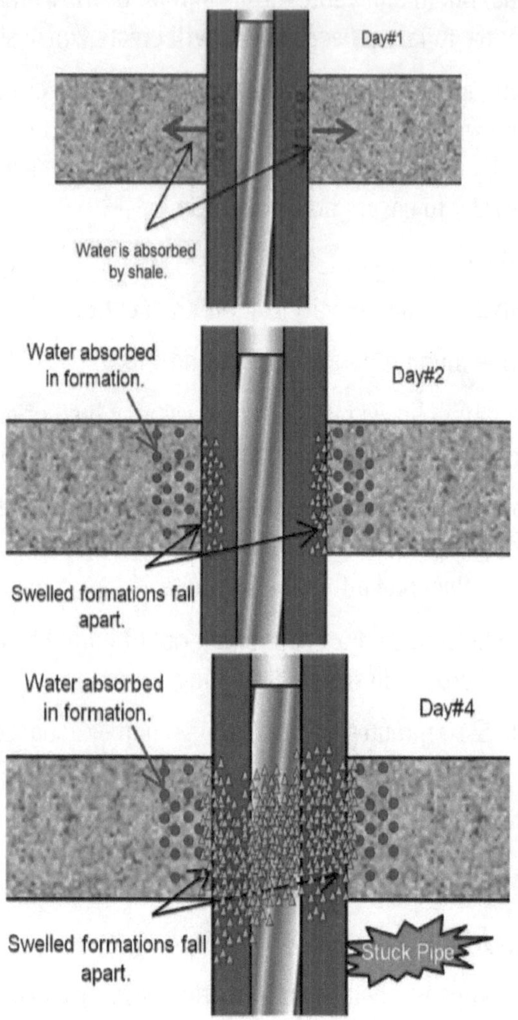

Warning signs of shale instability

- Torque and drag increase. An over pull may be observed.

- Mud properties became worse. You will see an increase in plastic viscosity, yield point (drilling mud becomes thicker).

- Pump pressure increases.

- Observe soft shale over shale shakers.

Indications when you stuck due to shale instability

· When it happens, you may observe very high pump pressure at small rate and sometimes circulation may be impossible.

· Most of the time it will happen when pulling out of hole. However, it can be possibly occurred while drilling as well.

What should you do for this situation?

1. Attempt to circulate with low pressure (300-400 psi). Do not use high pump pressure because the annulus will be packed harder and you will not be able to free the pipe anymore.

2. If you are drilling or POOH, apply maximum allowable torque and jar down with maximum trip load.

3. If you are tripping in hole, jar up with maximum trip load without applying any torque.

4. Attempt until pipe free and circulate to clean wellbore

Preventive actions:

1. For water based mud – you may need to add some salts that compatible with a mud formula in order to reduce chemical reaction between water and shale. Moreover, you should consider adding some coating polymers to prevent water contact with formation.

2. Use oil based mud instead of water based mud because oil will not react with shale.

3. Keep good flow rate to ensure good hole cleaning.

4. Perform back reaming and/or wiper trip.

5. Keep good mud properties.

Unconsolidated Formation Causes Stuck Pipe

How does it happen?

The situation could happen when drilling into unconsolidated formations such as gravel, sand, pea, etc. Since bond between particles are weak, particles in the formations will separate and fall down hole. If there are a lot of unconsolidated particles in the annulus, the drilling string can possibly be packed off.

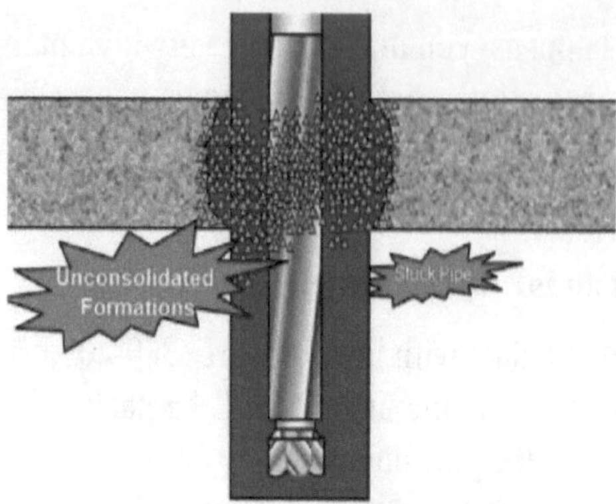

Warning signs when you get stuck due to unconsolidated formation

·	This situation could happen either while drilling or tripping. There is more chance that the situation can happen while drilling.

·	Slightly loss may possibly be seen while drilling.

·	Drilling torque and pump pressure abnormally increase.

·	Abnormal drag can be observed while picking up pipe.

Indications when you are stuck due to unconsolidated formation

·	Observe a lot of particles of gravel, sand, pea over shale shakers.

·	Increase in mud weight, rheology and sand content in drilling mud.

·	When it happens, the annulus may be completely packed off or bridged off; therefore, circulation is very difficult or impossible to establish.

·	Most of the time this situation happens while drilling a surface section where formation bonding is not strong. Moreover, it can occur suddenly.

What should you do for this situation?

1.	Attempt to circulate with low pressure (300-400 psi). Higher pump rate is not recommended because it will cause more cutting accumulation around a drill string and your drillstring will become harder to get free.

2. If you are drilling or POOH, apply maximum allowable torque and jar down with maximum trip load.

3. If you are tripping in hole, jar up with maximum trip load without applying any torque.

4. When the pipe is free, circulate to clean wellbore prior to drilling ahead.

Preventive actions:

1. Use high vis/weight sweep to help hole cleaning.

2. Ensure that fluid loss of drilling mud is not out of specification. Good fluid loss will create good mud cake which can help seal the unbounded formation.

3. Control ROP while drilling into unconsolidated zones and take time to clean the wellbore if necessary.

4. Slow tripping speed when BHA is being passed unconsolidated zones to minimize formation falling down.

5. Minimize surge pressure by starting/stopping pumps slowly and working string slowly.

6. Spot gel across suspected formations prior to tripping out of hole. Gel could prevent some particles to fall down into the wellbore.

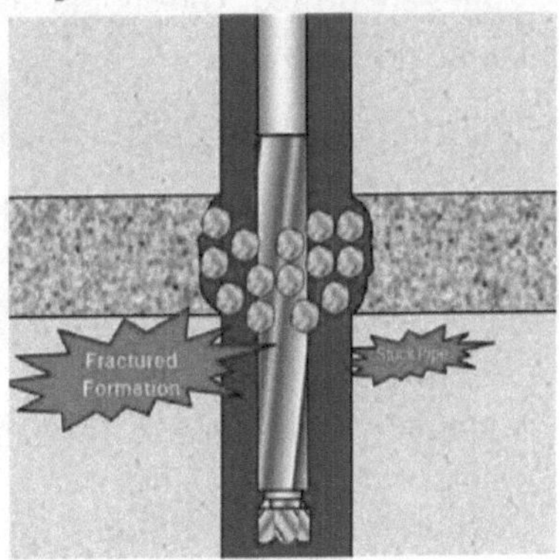

Stuck pipe by naturally fractured formation

Warning signs when you get stuck due to Fractured Formation

- Drill into potential naturally fractured zones as limestone, sand stone, carbonate, etc

- Observe big caving formations on shale shakers while drilling
- Observe volume to fill the hole is more than normal hole size

Stuck identification for Fractured Formation

- This situation can be occurred during drilling or tripping.
- Torque and drag are suddenly changed and erratic while drilling.
- Over pull off slip is noticed.
- Circulation could be restricted (you may get or not get good circulation)

What should you do for this situation?

1. Stuck while moving up, jar down with maximum allowable trip load without applying any torque!!!

2. Stuck while moving down, jar up without apply torque

3. Pump weighted hi-vis sweep with maximum allowable flow rate

Preventive actions:

1. Keep mud in good shape. Good and thin mud cake can support fracture formation in some cases.

2. If the suspected zones are drilled, you should take time to circulate hole clean before making head way.

3. Start and stop circulation slowly to minimize surge pressure.

4. Work pipe with restricted speed to prevent surging formations.

5. Tripping speed should be slow while BHA is being run into suspected zones.

6. The fractured formations require time to get stabilized.

Cement Blocks Causes Stuck Pipe

How does it happen?

Cement around casing shoe or open hole cement squeeze becomes unstable and finally chunks of cement fall into a wellbore. If there are a lot of cement chunks in the annulus, drilling string will be stuck.

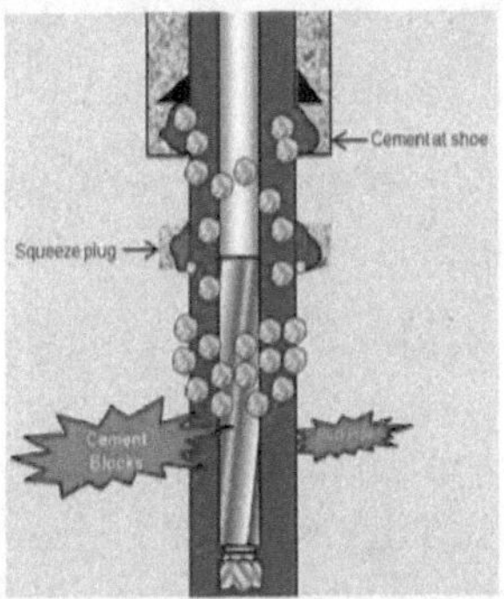

Cement Blocks

Warning signs when you get stuck due to Cement Blocks

- Rat hole is too long.

- Drilling in to areas where open hole cement jobs as cement squeeze or kick off plug were performed.

Stuck identification for Cement Blocks

- Cement chunks are seen at shale shakers.

- There are cement content in mud logger samples.

- Stuck pipe due to cement blocks can be occurred anytime.

- Circulation is not restricted.

- Torque and drag are drastically increased and erratic.

What should you do for this situation?

1. Stuck while moving up, jar down with maximum allowable trip load. Gradually apply torque if required.

2. Stuck while moving down, jar up without applying torque.

3. Pump weighted hi-vis sweep with maximum allowable flow rate to clean large pieces of cement around drilling string/BHA.

Preventive actions:

1. Do not leave a long rat hole.

2. Ream with circulation through casing shoe and areas where there is open hole cement.

3. Attempt to clean cement in the annulus prior to drilling.

4. Wait for cement setting long enough before drilling ahead.

5. Minimize tripping speed when BHA passes through casing shoe or cement plug/cement squeeze depth.

Soft Cement Causes Stuck Pipe

How does it happen?

The drillstring/BHA is in soft cement and when circulation is established, pumping pressure causes the soft cement to flash set (cement becomes harder quickly). Finally the drill string gets stuck due to hard cement around it.

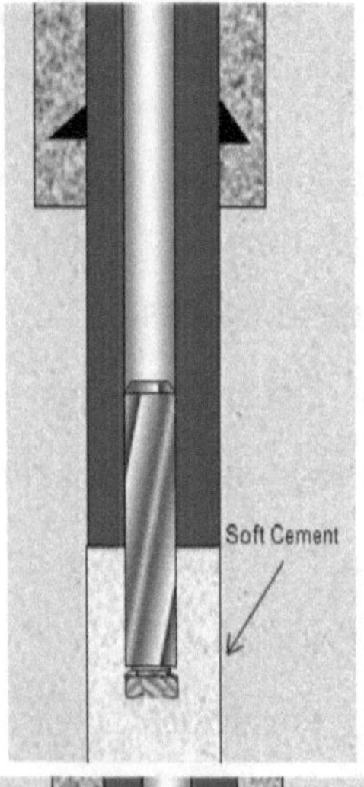

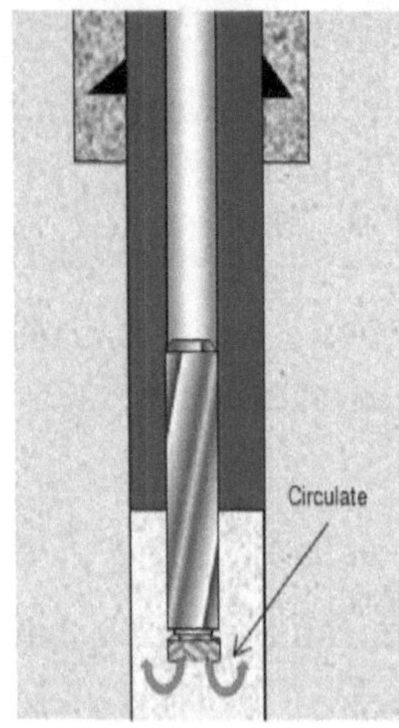

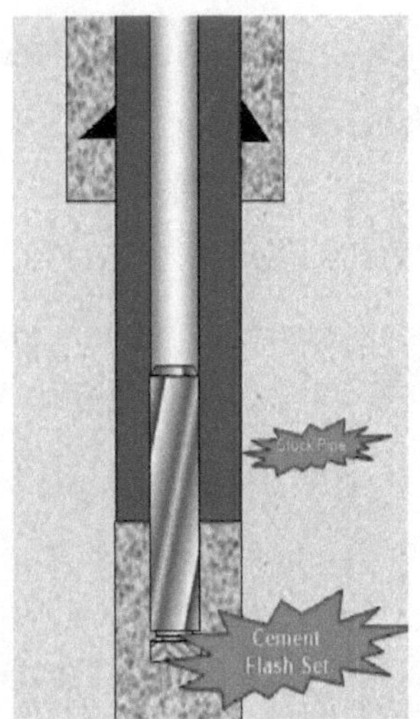

Warning signs when you get stuck due to Cement Blocks

· Run in hole after the open hole cement job as cement balanced plug is completed.

· Unable to see firm cement while attempting to find the theoretical top of cement. It indicates that you may be in the soft cement.

Stuck identification for Soft Cement

· It happens when pump pressure is brought up and pump pressure increases quickly.

· Rotary torque suddenly increases.

· When the soft cement is flash set, you may not be able to get circulation or get low circulation at very high pump pressure.

What should you do for this situation?

1. First of all, before jarring operation, you must bleed off trapped pressure in the string.

2. Apply jar with maximum trip load. Jar at the opposite direction of string movement. For example, if you are stuck while moving up, you need to jar down. On the other hand, you need to jar up, if you are stuck while moving down.

Preventive actions:

1. Ensure that cement is properly set prior to tripping to top of cement.

2. Stop at least 100 ft above the calculated top of cement and establish circulation prior to tag top of cement.

3. Tag cement slow with pump on.

4. Don't clean out cement too fast. Attempt to control drill and check pick up/slack off weight and torque frequently while drilling out cement.

Junk Causes Stuck Pipe

How does it happen?

Junk from the surface drops into the wellbore casing stuck pipe. It could be happened due to several factors as poor housekeeping on the rig floor, rotary table not covered, surface/down hole equipment failure.

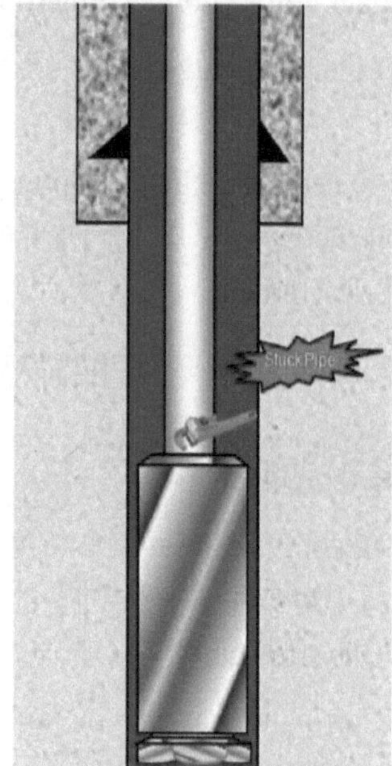

Warning signs when you get stuck due to junk

- Observe equipment on surface falling downhole

- If down hole equipment failure, the drill string gets jammed suddenly without any sings.

- Stuck pipe by the junk can be occurred any time.

- Suspicious substance as metal, wood, rubber may be found at the shale shaker.

Stuck identification for Junk

- Torque suddenly becomes erratic.

- Drag increases

- Equipment on the rig floor falls down hole.

What should you do for this situation?

1. If you get stuck while moving up, jar down with maximum trip load. Torque may be applied with caution.

2. If you get stuck while moving down, jar up without any toque applied in the drill string.

Preventive actions:

1. Maintain good housekeeping on the rig floor

2. Ensure that hole cover is used all the time when work on rotary table.

3. Maintain tool used on the rig floor in a good condition.

4. Inspect downhole tool prior to tripping in hole.

How to Free Stuck Pipe Caused by Pack off / Bridging

If you already know that the stuck pipe is caused by wellbore geometry, these following instructions are guide lines on how to free the stuck drill string.

What should you do to free the stuck pipe caused by Pack off / Bridging?

- Circulate with low flow rate (300 – 400 psi pumping pressure). This is very important to apply low flow rate because if high flow rate is applied, the stuck situation becomes worse.

- If the drill string gets stuck while moving up or with the string in static condition, jar down with maximum trip load and torque can be applied into drill string while jarring down. DO NOT JAR UP. Be caution while applying torque, do not exceed make up torque.

- On the other hand, if the drill string gets stuck while moving down, jar up with maximum trip load. <u>DO NOT apply torque in the drill string while jarring up</u>.

·   To free the string, jarring operation may take long time (10 hours +) so please be patient.

What should you do after the string becomes free?

·   Increase flow rate and circulate to clean wellbore at maximum allowable flow rate. Flow rate must be more than cutting slip velocity in order to transport cuttings effectively.

·   Reciprocate and rotate while circulating to improve hole cleaning ability. Work the drill string with full stand if possible.

·   Ensure that the wellbore is clean prior to continuing the operation. You can see from the sale shaker whether the hole is clean or not.

·   Sweep may be utilized to improve hole cleaing.

·   Back ream or make a short trip through the area where causes the stuck pipe issue.

Stuck Pipe Caused By Differential Sticking

How does it happen?

Differential sticking is one of the most common causes of pipe stuck. It can happen when there is differential pressure (overbalance pressure) pushing a drillstring into filter cake of a permeable formation.

How does it happen?

Differential sticking is one of the most common causes of pipe stuck. It can happen when there is differential pressure (overbalance pressure) pushing a drillstring into filter cake of a permeable formation.

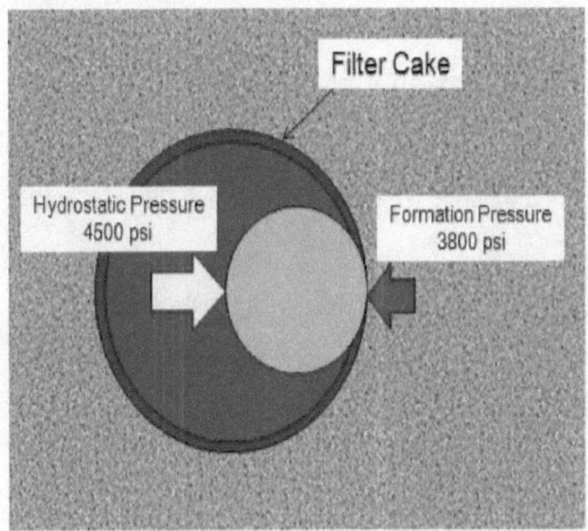

<u>Four factors causing the differential sticking are as follows:</u>

Permeable formation as sand stone, lime, carbonate, etc.

Overbalance – typically mud weight in the well is more than formation pressure. More overbalance in the wellbore, more chance of getting differential sticking.

Filter cake – Poor and thick filter cake increases chances of sticking the drill string.

Pipe movement – if the drillstring is station for a period of time, the filter cake will tend to develop around permeable zones and the drillstring. Therefore, potential of getting differentially stuck is increased.

Warning signs when you get stuck due to differential sticking

·   There are high over balance between wellbore and formation. Especially, when there is highly depleted formation, the chance of getting differentially stuck is so high.

·   Torque, pick up and slack off weight increase when the drill string is being moved. Once it happens, you may not be able to pull or rotate pipe.

Stuck identification for differential sticking

·   Drill string is in station for a period of time. The differential sticking is happened when there is no pipe movement for long time.

·   Circulation can be established without increasing in pressure.

- BHA is across the permeable zone.

Let's see how much differential force from this situation

- Formation pressure = 3800 psi

- Hydrostatic pressure =4500 psi

- Cross area of stuck pipe = 1500 square inch

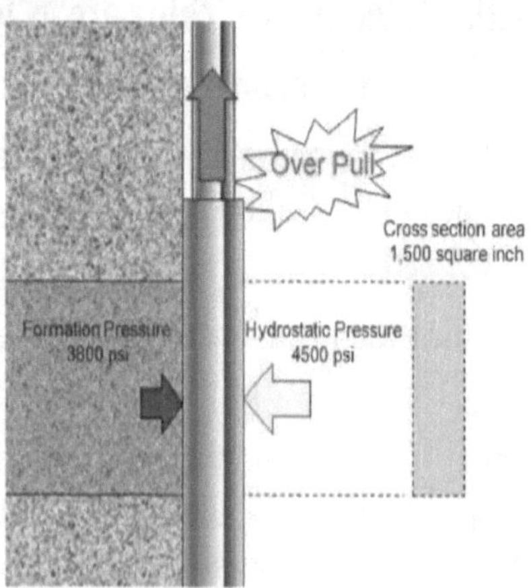

You can determine how much differential force based on a **following formula:**

Force = Differential Pressure x Cross Section Area

Where;

- Force is in lb.

- Differential pressure is in psi.

- Cross section area is in square inch.

- Force = (4500 – 3800) x 1500

- Force = 1,050,000 lb

This is massive !!!

If we assume a coefficient friction of 0.5, you can determine how much tension you need to free the pipe.

From the basic of physic,

F= coefficient friction x N

Where;

- F is force to pull.

- N is reactive force.

- For this scenario, N is equal to differential force.

- F = 0.5 x 1,050,000 = 525,000 lb

You need overpull of 525,000 lb to free the pipe from this situation.

What should you do for this situation?

1. Apply torque into drill string and jar down with maximum allowable trip load

2. Jar up without apply torque

3. Spot light weight pill to decrease hydrostatic pressure. If you want to the light weight pill, you must ensure that the overall hydrostatic pressure is more than formation pressure. Otherwise, you will face with a well control situation.

Preventive actions:

1. Do not use too high mud weight

2. Do not stop moving string for a period of time, especially, when the BHA is across formations.

3. Keep mud in good shape. Under specification drilling mud will create thick mud cake which can be a big impact for the differential sticking.

4. Minimize length of BHA and use spiral drill collar and heavy weight drill pipe to reduce contact area.

How To Free Stuck Pipe Caused By Differential Sticking

These following guidelines help you free stuck drillstring caused by differential sticking.

The first action that you should do to free the stuck pipe caused by differential sticking.

- Apply maximum flow rate as much as you can.

- Apply maximum torque in the drillstring and work down torque to stuck depth. Torque in the string will improve chance of free the pipe.

- Slack off weight of string to maximum sit down weight.

- Jar down with maximum trip load. Torque may be applied with jarring down with caution. The chance of freeing the pipe by jarring down is more than jarring up. Please be patient when a hydraulic jar trips because it may take around 5 minutes each circle.

The secondary action to free the pipe that you may try

- Reduce hydrostatic pressure by pumping low weight mud/pill. You must ensure that overall hydrostatic pressure is still able to control reservoir fluid to accidentally come into the wellbore.

- Continue jarring down with maximum trip load and apply torque into drill string.

- It may take long time to free the pipe therefore personnel must be patient.

What should you do after the string becomes free?

- Circulate at maximum allowable flow rate. Flow rate must be more than cutting slip velocity in order to transport cuttings effectively.

- Reciprocate and work pipe while cleaning the hole. Ensure that you can work pipe with full stand or joint while circulating.

- Condition mud prior to drilling ahead because if you still drill with poor mud properties, the differential sticking will be re-occurred.

Stuck Pipe Caused By Wellbore Geometry

There are eleven cases in this category.

1. Stiff bottom hole assembly
2. Key seat
3. Micro doglegs
4. Ledges
5. Mobile formations
6. Undergauge hole
7. Hydro-Pressured Shale
8. Geo-Pressured Shale
9. Overburden Stress
10. Tectonic Stress
11. Unconsolidated Formation

Stiff BHA Causes Stuck Pipe

How does it happen?

The well is drilled with limber BHA. When the limber BHA is pulled out and the stiffer BHA is used as the next BHA, the stiff BHA is unable to pass the existing hole due to excessive dog leg and finally the BHA gets stuck.

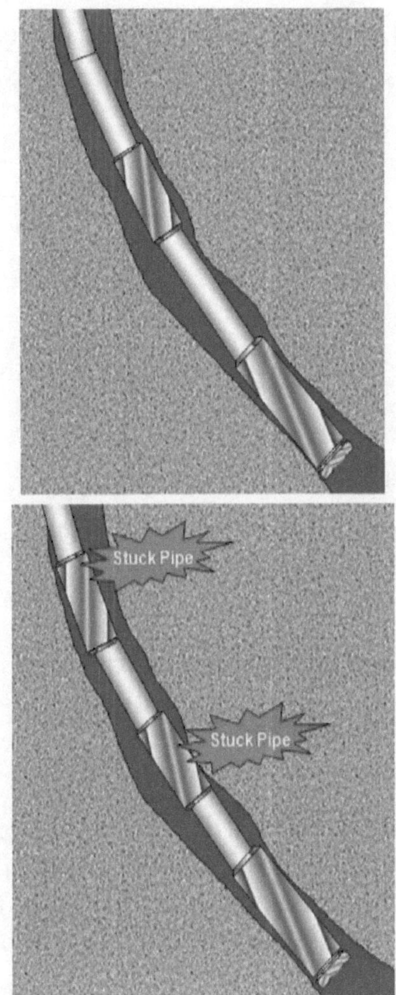

Warning signs when you get stuck due to Stiff BHA

- Excessive dog leg severity is drilled.

- The pulled BHA is under gauge.

- While tripping in hole with new BHA, sudden sit down weight is seen.

Stuck identification for Stiff BHA

- The BHA has a possibility to get stuck at high dog leg areas.

- It is most likely occurred while running in hole.

- Circulation is not restricted.

What should you do for this situation?

1. If you get stuck while moving up, jar down with maximum trip load. Torque may be applied with caution.

2. If you get stuck while moving down, jar up without any toque applied in the drill string.

Preventive actions:

1. Do not drill with a lot of dog leg severity

2. Minimize changes in BHA configuration

3. If the stiffer is used, you may need to ream in hole passing through high dog leg areas. Moreover, tripping/reaming speed must be limited prior to entering high risk zones.

4. Do not create a lot of sit down weight to minimize BHA jammed.

Key Seat Causes Stuck Pipe

How does it happen?

While drilling, with high tension and torsion in a drilling string, the drill string creates wear, called "key seat", at wellbore where there are changes in direction. The soft to medium hard formation has a great tendency to get key seat. While pulling out of the hole, BHA gets stuck into the key seat.

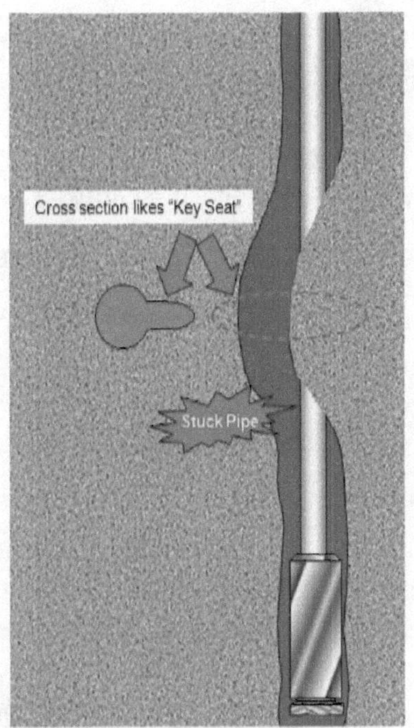

Warning signs when you get stuck due to Key Seat

- High dog leg severity

- Long drilling hours without reaming back through the high dog leg area

Stuck identification for Key Seat

· This situation is occurred when pulling out of hole only.

· Circulation is not restricted.

· High over pull is suddenly seen when the BHA is pulled into the key seat.

· Tripping back is possible.

What should you do for this situation?

1. Because the drill sting gets stuck while moving up, jar down with maximum trip load must be applied. Torque while jarring down can be applied as well.

2. Bring the rotation at slow speed and attempt to ream back with small over pull into the key seat areas.

Preventive actions:

1. Do not drill with a lot of dog leg severity

2. Back ream some areas where high dog leg severity is presented.

3. Run short trip or wiper trip to minimize the key seat.

4. Utilize a reamer to high dog leg zones.

Micro Dogleg Causes Stuck Pipe

How does it happen?

Micro dogleg is occurred in areas where there are several corrections in inclination and azimuth and it most likely happens in hard/soft interbedd rock. If there are micro dogleg areas in the well, the bottom hole assembly can get stuck.

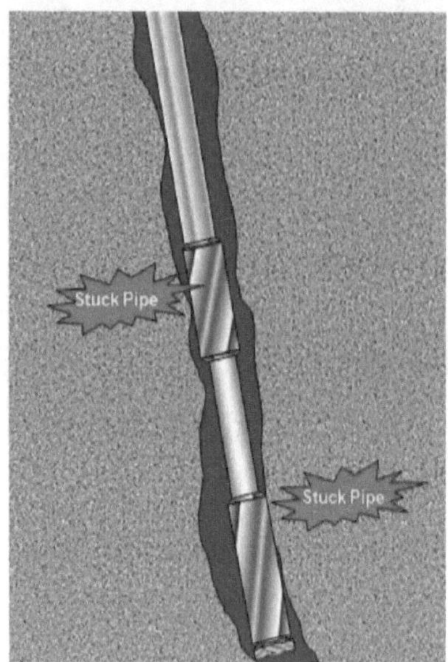

Warning signs when you get stuck due to Micro Dogleg

·   Hard and soft streak formations are drilled. You can easily observe from changes in ROP.

·   Inclination and azimuth are frequently changed.

- Drilling the well with a mud motor causes this issue because of rotating and sliding operation.

Stuck identification for Micro Dogleg

- Drilling torque and drag are erratic.

- It can be happened while tripping or drilling.

- Circulation can be established without any restriction.

What should you do for this situation?

1. If the drillstring is stuck while moving up, jar down with maximum trip load. Torque can be applied with caution while jarring down.

2. If the drillstring is stuck while moving down, jar up with maximum trip load without applying any torque in drill string.

3. If the drill string is free, you may need to consider back reaming to clear micro dogleg.

Preventive actions:

1. Minimize changes in inclination and azimuth

2. Back ream operation should be performed when hard/soft streak is drilled.

3. Slow down tripping sleep when entering possible problematic zones.

Ledges Cause Stuck Pipe

How does it happen?

Ledges are occurred while drilling in sequential formations which have soft, hard formations, and naturally fractured formations. Stabilizers in BHA and tool joint easily wear soft formations and naturally fractured formations, however, the hard formations are still in gauge (hole size not change). If there are a lot of ledges in the wellbore, the drillstring can get stuck under ledges.

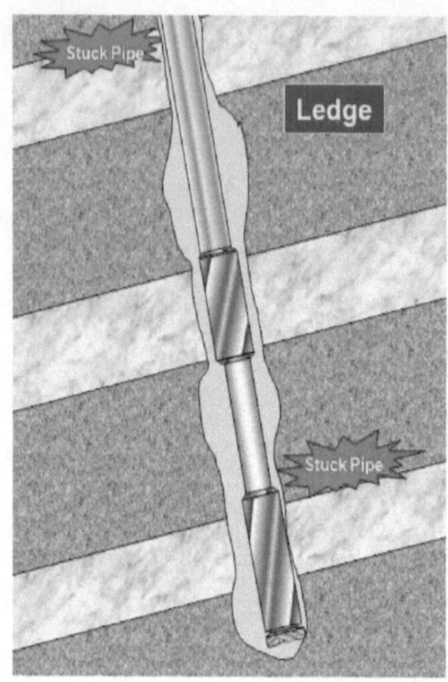

Warning signs when you get stuck due to Ledges

·   Hard and soft streak formations are drilled. You can easily observe from changes in ROP.

·   Mud logging samples show soft and hard rocks.

·   There is potential for fractured formations to be drilled.

Stuck identification for Ledges

·   Erratic over pull is observed.

·   It can be happened while tripping or drilling and it is also related to micro doglegs.

·   Circulation can be established without any restriction.

What should you do for this situation?

1. If the drillstring is stuck while moving up, jar down with maximum trip load. Torque can be applied with caution while jarring down.

2. If the drillstring is stuck while moving down, jar up with maximum trip load without applying any torque in drill string.

3. If the drill string is free, you may need to consider back reaming to

clear some ledges.

Preventive actions:

1. Minimize big stabilizers.

2. Minimize changes in inclination and azimuth.

3. Back reaming operation should be performed when the suspected formations are drilled. Carefully watch the over pull while reaming.

4. Slow down tripping sleep when entering possible problematic formations.

Mobile Formation Causes Stuck Pipe

How does it happen?

Mobile formation is caused by over burden pressure that squeezes shale and/or salt into a wellbore. The squeezed formations reduce wellbore diameter; therefore, the drillstring/BHA gets stuck due to under gauge wellbore.

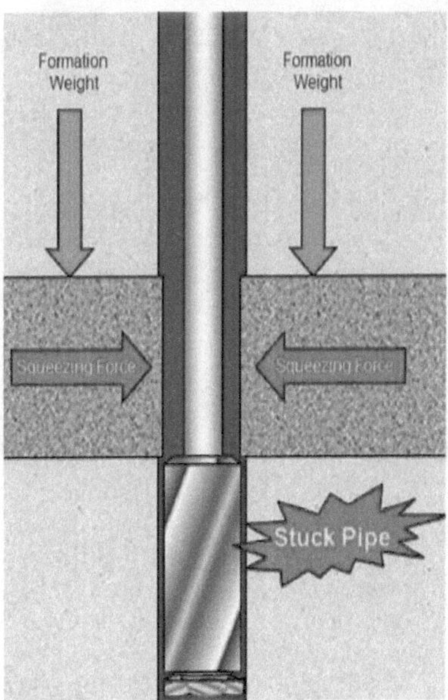

Warning signs when you get stuck due to Ledges

·   Salt and shale are drilled. You can see from the Mud logging samples showing shale. Moreover, Chloride content must be increase in case of drilling into salt zones.

Stuck identification for Ledges

· Over pull, down weight and torque are suddenly increased.

· It could be happened anytime as drilling, tripping in and tripping out depending on how fast plastic formations are moved.

· Most of the time, the BHA gets stuck at the plastic zones because BHA contains the largest diameter component.

· Circulation is not restricted or just slightly restricted.

What should you do for this situation?

1. If the drillstring is stuck while moving up, jar down with maximum trip load. Torque can be applied with caution while jarring down.

2. If the drillstring is stuck while moving down, jar up with maximum trip load without applying any torque in drill string.

3. If you are sure that the plastic formations drilled are salt, you may consider spotting fresh water to dissolve the salt. However, you need to consider regarding well control issue.

Preventive actions:

1. Use sufficient mud weight to hold back formation movement.

2. Back ream and wiper trip should be performed across the suspected rocks.

3. Minimize time for open hole exposure.

4. Trip in hole with caution prior to entering possible problematic formations.

Undergauge Hole Causes Stuck Pipe

How does it happen?

Undergauge hole can be happened when drilling in to hard and abrasive formations where wears a drill bit. When the bit is undergauge because the abrasive formation wears a bit and stablizers, a hole size becomes smaller. When the new BHA is run in hole, the new bit/BHA gets stuck into the undergauge hole section. Additionally, if coring operation is performed with smaller core bit than the next bit, the new bit can get stuck at the top of coring section.

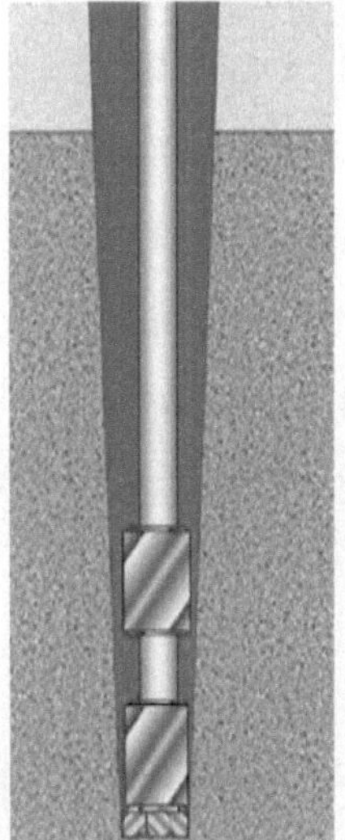

Hole becomes smaller due to bit and stabilizers worn out

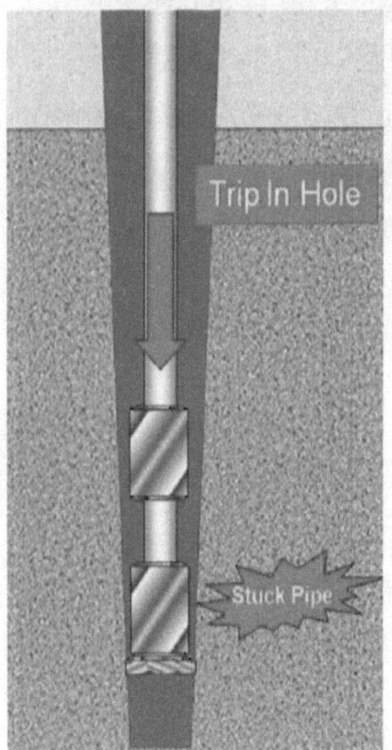

The next BHA gets stuck.

Warning signs when you get stuck due to Undergauge Hole

- Drilling into abrasive formations.

- A bit and stabilizers are undergauge.

Stuck identification for Undergauge Hole

- This type of stuck pipe is occurred only when tripping in hole.

- Sit down weight suddenly increases.

- The bit gets jams off bottom.

- Circulation is able to be established.

What should you do for this situation?

This stuck pipe is always happen while the drillstring is being moved down; therefore you need to jar up with maximum trip load without applying any torque in drill string.

Preventive actions:

1. Properly gauge bit/stabilizer after pulled out so you will know the possibility right away.

2. Do not stag weight in order to pass the tight spots. The more weight you put on top, the harder to free the pipe.

3. If the undergauge bit/stabilizer is observed, you need to ream down at least 1-2 stands off bottom.

4. Reaming at least 1-2 stands above top of coring section.

5. Trip in hole with controlled speed prior to going to possible problematic areas.

Hydro-Pressured Shale Causes Stuck Pipe

How does it happen?

Hydro-pressured shale is a common problem in some area and it could cause stuck pipe.With mud weight in the wellbore higher than formation pressure, pore pressure of shale is always charged by hydrostatic pressure from drilling mud.

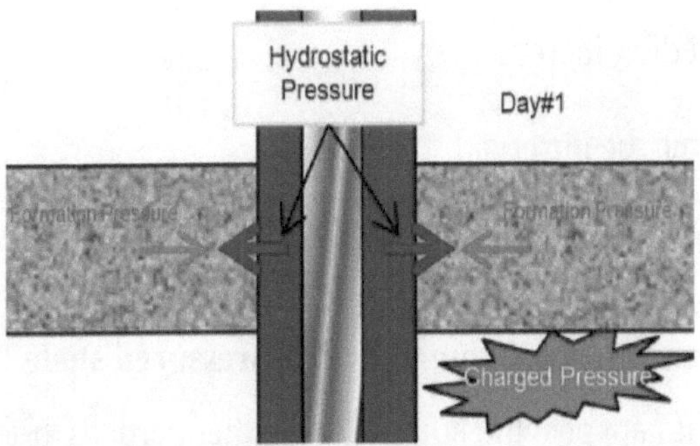

When the well has been drilled for a period of time, shale formations become unstable due to charged pressure and finally shale breaks apart and falls down into the hole.

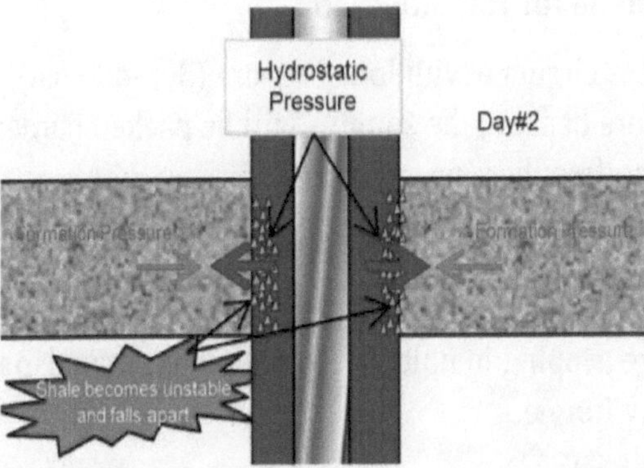

Finally, a drill string gets stuck due to hydro-pressured shale which accumulates in the annulus.

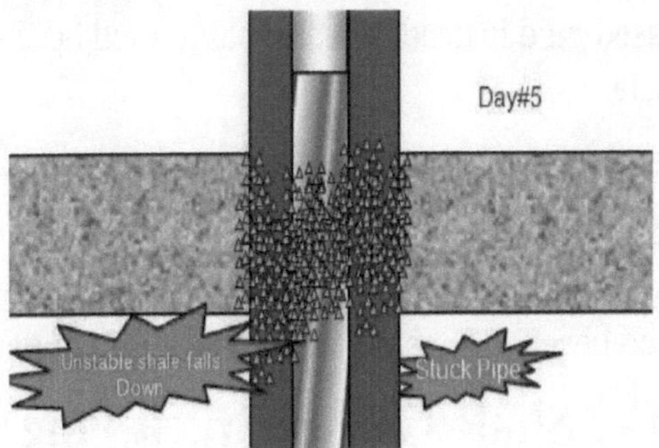

This process is time dependent like shale instability. It may take days before the stuck pipe situation will be occurred.

Warning signs of hydro-pressured shale

- Torque and drag increase.

- Over pull may be observed.

- Observe shale caving on shale shakers

Indications when you stuck due to hydro-pressured shale

· When it happens, the hole will be either partially bridge off or packed off; therefore, circulate is restricted or impossible in some cases.

· It could be happened while tripping and drilling.

What should you do for this situation?

1. Attempt to circulate with low pressure (300-400 psi). Do not use high pump pressure because the annulus will be packed harder and you will not be able to free the pipe anymore.

2. If you are drilling or POOH, apply maximum allowable torque and jar down with maximum trip load.

3. If you are tripping in hole, jar up with maximum trip load without applying any torque.

4. Attempt until pipe free and circulate to clean wellbore.

Preventive actions:

1. Use oil based mud instead of water based mud because oil will not react with shale.

2. Minimize surge pressure and equivalent circulating density (ECD) in the wellbore.

3. Keep mud properties in good shape. Avoid drilling and circulating with thick mud because it creates additional surge pressure.

Geo-Pressured Shale Causes Stuck Pipe

How does it happen?

Pore pressure in shale is more than hydrostatic pressure; however the well does not flow because shale is in permeable. While drilling through pressured shale formation, pressure in shale causes fractures of shale due to stress crack.

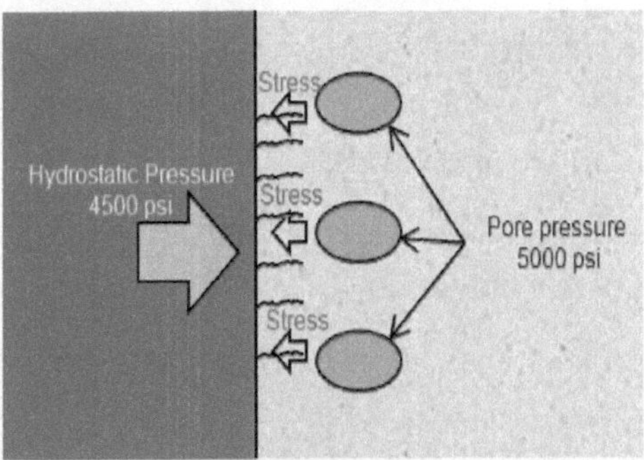

Shale finally falls into the well and results in stuck pipe incident.

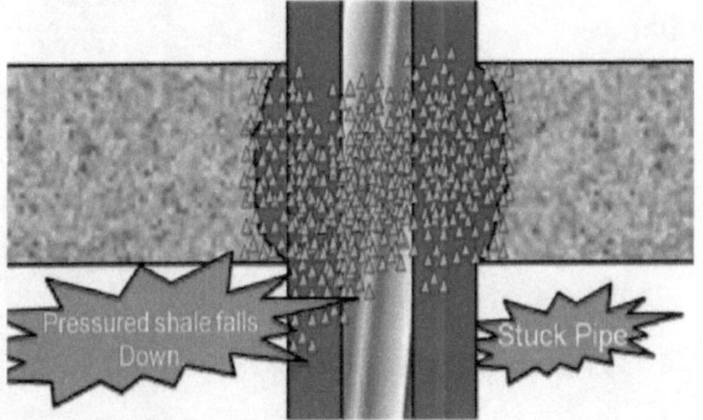

Warning signs of geo-pressured shale

- Shale fractures are seen at shale shakers.

- Possibly observe change in d-exponent, which means a sign of abnormal pressure, from mud logger.

- Pump pressure abnormally increases.

- Increase in rate of penetration (ROP).

- When compared to a normal trend, torque and drag trend abnormally increase.

- Background gas may increase.

Indications when you stuck due to geo-pressured shale

1. It could be happened either while tripping or drilling.

2. When it happens, the hole may be completely packed off; therefore, circulate is restricted or impossible in some cases.

What should you do for this situation?

1. Attempt to circulate with low pressure (300-400 psi). Do not use high pump pressure because the annulus will be packed harder and you will not be able to free the pipe anymore.

2. If you are drilling or POOH, apply maximum allowable torque and jar down with maximum trip load.

3. If you are tripping in hole, jar up with maximum trip load without applying any torque.

4. Attempt until pipe free and circulate to clean wellbore.

Preventive actions:

1. Use proper mud weight to create over balance. You may need to weight up prior to drilling in to high pressure shale zones.

2. Minimize surge pressure and equivalent circulating density (ECD) in the wellbore

Overburden Stress Shale Causes Stuck Pipe

How does it happen?

Overburden stress increases over depth (the more a well is drilled, the more overburden stress will be seen). When mud weight is not enough to support the overburden, the stress from the overburden will create shale fractures which will fall down into the wellbore. Finally, shale fractures will pack the wellbore and cause a stuck pipe incident,

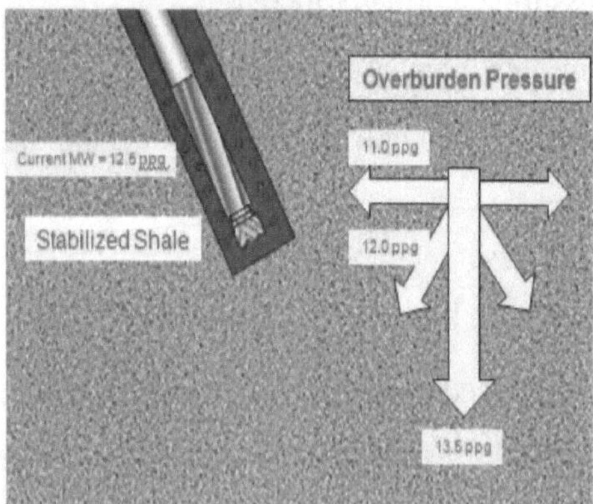

(Mud weight is high enough to overcome overburden stress.)

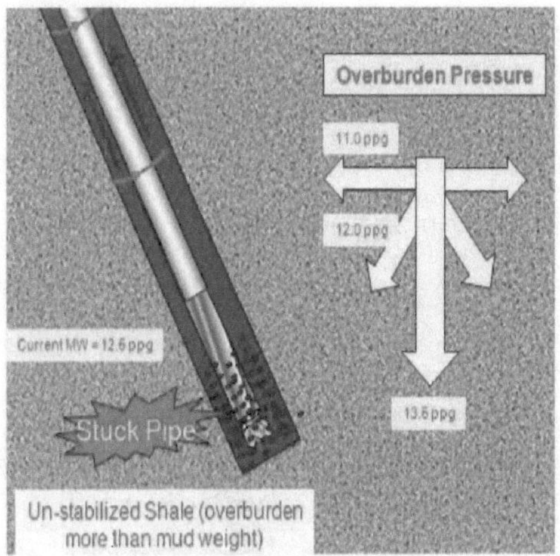

Mud weight is not high enough to overcome overburden stress.)

Warning signs of overburden stress shale

- Torque and drag increase.

- Pump pressure increase.

- Abnormal amount of shale at shale shakers

- Caving shape of shale at shakers

Indications when you stuck due to overburden stress shale

- It could be happened either while tripping or drilling (most likely while drilling).

- When it happens, the hole may be completely packed off or bridged off; therefore, circulation is very difficult or impossible to establish.

What should you do for this situation?

1. Attempt to circulate with low pressure (300-400 psi). Do not use high pump pressure because the annulus will be packed harder and you will not be able to free the pipe anymore.

2. If you are drilling or POOH, apply maximum allowable torque and jar down with maximum trip load.

3. If you are tripping in hole, jar up with maximum trip load without applying any torque.

4. Attempt until pipe free and circulate to clean wellbore.

Preventive actions:

1. Use drilling mud that heavy enough to stabilize overburden stress.

2. Weight up mud prior to drilling into stressed shale zones.

Tectonic Stress Causes Stuck Pipe

How does it happen?

Tectonic stress is a nature phenomenal and it naturally occurs due to lateral force from the formation. Typically, if the rig is close to mountains, there is high chance to face with the tectonic stress issue.

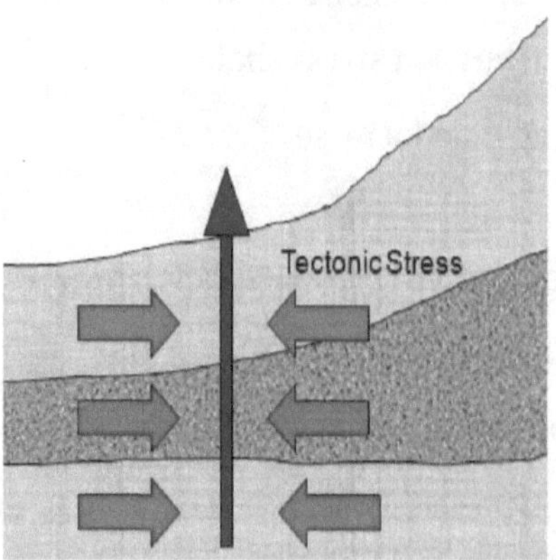

The later force will create stress which will squeeze sandstone causing under gauge hole. Moreover, the lateral stress will fracture shale and create additional cuttings.

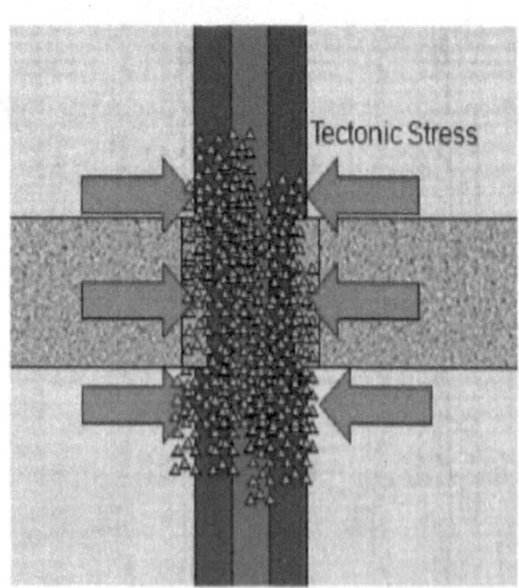

These two main results (under gauge in sand stone and shale fractures) can cause stuck pipe situations.

Warning signs when you get stuck due to tectonic stress

- The rig location is nearby mountains.

- Drilling torque is erratic.

- Abnormal drag is happened.

- Shale fractures are observed at shale shakers.

Indications when you are stuck due to tectonic stress

- It could be happened either while tripping or drilling.

- When it happens, the hole may be completely packed off or bridged off; therefore, circulation is very difficult or impossible to establish.

What should you do for this situation?

1. Attempt to circulate with low pressure (300-400 psi). Higher pump rate is not recommended because it will cause more cutting accumulation around a drill string.

2. If you are drilling or POOH, apply maximum allowable torque and jar down with maximum trip load.

3. If you are tripping in hole, jar up with maximum trip load without applying any torque.

4. Attempt until pipe free and circulate to clean wellbore.

Preventive actions:

1. Increase mud weight to increase wellbore stability.

2. Use high vis/weight sweep to help hole cleaning.

3. Minimize operation time. The less time you drill and complete the well, the less tectonic stress affects your wellbore.

Unconsolidated Formation Causes Stuck Pipe

How does it happen?

The situation could happen when drilling into unconsolidated formations such as gravel, sand, pea, etc. Since bond between particles are weak, particles in the formations will separate and fall down hole. If there are a lot of unconsolidated particles in the annulus, the drilling string can possibly be packed off.

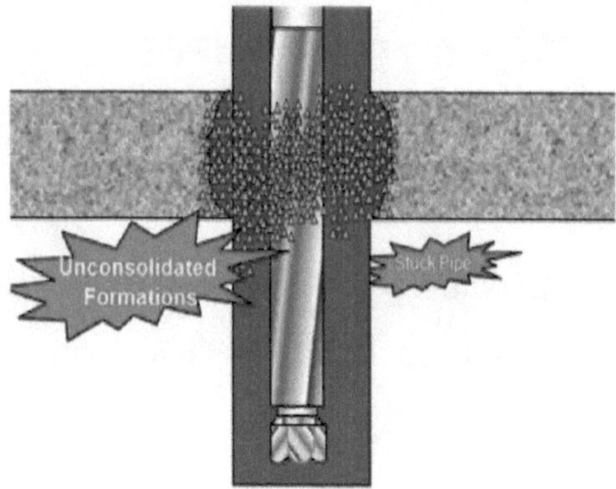

Warning signs when you get stuck due to unconsolidated formation

- This situation could happen either while drilling or tripping. There is more chance that the situation can happen while drilling.

- Slightly loss may possibly be seen while drilling.

- Drilling torque and pump pressure abnormally increase.

- Abnormal drag can be observed while picking up pipe.

Indications when you are stuck due to unconsolidated formation

- Observe a lot of particles of gravel, sand, pea over shale shakers.

- Increase in mud weight, rheology and sand content in drilling mud.

- When it happens, the annulus may be completely packed off or bridged off; therefore, circulation is very difficult or impossible to establish.

- Most of the time this situation happens while drilling a surface section where formation bonding is not strong. Moreover, it can occur suddenly.

What should you do for this situation?

1. Attempt to circulate with low pressure (300-400 psi). Higher pump rate is not recommended because it will cause more cutting accumulation around a drill string and your drillstring will become harder to get free.

2. If you are drilling or POOH, apply maximum allowable torque and jar down with maximum trip load.

3. If you are tripping in hole, jar up with maximum trip load without applying any torque.

4. When the pipe is free, circulate to clean wellbore prior to drilling ahead.

Preventive actions:

1. Use high vis/weight sweep to help hole cleaning.

2. Ensure that fluid loss of drilling mud is not out of specification. Good

fluid loss will create good mud cake which can help seal the unbounded formation.

3. Control ROP while drilling into unconsolidated zones and take time to clean the wellbore if necessary.

4. Slow tripping speed when BHA is being passed unconsolidated zones to minimize formation falling down.

5. Minimize surge pressure by starting/stopping pumps slowly and working string slowly.

6. Spot gel across suspected formations prior to tripping out of hole. Gel could prevent some particles to fall down into the wellbore.

How To Free Stuck Pipe Caused By Wellbore Geometry

If you already know that the stuck pipe is caused by wellbore geometry, these following instructions are guide lines on how to free the stuck drill string.

What should you do to free the stuck pipe caused by wellbore geometry?

- If the drill string gets stuck while moving up, jar down with maximum trip load and torque can be applied into drill string while jarring down. Be caution while applying torque, do not exceed make up torque.

- On the other hand, if the drill string gets stuck while moving down, jar up with maximum trip load. DO NOT apply torque in the drill string while jarring up.

- Flow rate must be reduced while jarring up/down. Do not use high flow rate because it will make the stuck situation became worse and you will not be able to free the pipe forever.

- To free the string, jarring operation may take long time so please be patient.

- If a formation you get stuck is limestone or chalk, acid can be spotted to dissolve cuttings around the pipe.

- If the drill string is stuck in a salt formation, spotting fresh water is another choice to clear the salt in the annulus.

- Please always seriously consider regarding well control prior to spotting light weight stuff (acid or fresh water) around the drill string. You must ensure that you are still over balance formation pressure otherwise you will be dealing with well control too.

What should you do after the string becomes free?

- Increase flow rate and circulate to clean wellbore. Flow rate must be more than cutting slip velocity in order to transport cuttings effectively.

- Reciprocate and work pipe while cleaning the hole.

- Ensure that the wellbore is clean prior to continuing the operation.

- Back ream or make a short trip the section that causes the problem

How To Determine Stuck Depth

First Method of Stuck Depth Calculation

The steps of stuck depth calculation are as follows:

1) Determine the free point constant (FPC) by this following formula:

FPC = As x 2500

where: As = pipe wall cross sectional area, sq in.

$$As = (OD^2 - ID^2) \times 0.7854$$

OD and ID are inch unit.

2) Determine depth of stuck pipe by this following formula:

Depth of stuck pipe = (Pipe stretch in inch x free point constant (FPC)) ÷ Pull force in thousands of pounds

Please see the example below to demonstrate you how to apply those 2 formulas above into real drilling operation.

Example: Determine the free point constant (FPC) and the depth the pipe is stuck using the following information:

- 3-1/2" tubing # 9.5 lb/ft

- 3-1/2" tubing ID = 2.992 inch

- 20 inch of stretch with 25 Klb of pulling force

1) Determine free point constant (FPC):

- $FPC = (3.5^2 - 2.992^2) \times 0.7854 \times 2500$

- FPC = 6475.5

2) Determine the depth of stuck pipe:

- Stuck depth of pipe (ft) = (20 inch x 6475.5) ÷ 25 klb

- Stuck depth of pipe (ft) = 5,180 ft

Second Method of Stuck Depth Calculation

This is another formula for determining the stuck depth.

Stuck depth in ft = (735,294 x e x Wdp) ÷ (Stretch pull, lb)

$$Wdp = 2.67 \times (OD^2 - ID^2)$$

Where;

e = pipe stretch, in.

Wdp = drill pipe weight or tubular, lb/ft (plain end)

Example: Determine the free point constant (FPC) and the depth the pipe is stuck using the following information:

- 3-1/2" tubing # 9.5 lb/ft

- 3-1/2" tubing ID = 2.992 inch

- 20 inch of stretch with 25 Klb of pulling force

- Wdp = 2.67 x (3.52 – 2.9922)

- Wdp = 8.805

- Stuck depth = (735,294 x 20 x 8.805) ÷ (25,000)

- Stuck depth = 5,179 ft

- You can see that both calculation methods gives you the same stuck depth.

Table of Contents

9 798529 873205